YOUR KNOWLEDGE HAS VALUE

- We will publish your bachelor's and master's thesis, essays and papers

- Your own eBook and book - sold worldwide in all relevant shops

- Earn money with each sale

Upload your text at www.GRIN.com
and publish for free

Debajyoti Bose, Hiranmay Gangopadhyay

Effect of carbohydrates and amino acids on fermentative production of alpha amylase

Solid state fermentation utilizing agricultural wastes

GRIN Verlag

Bibliografische Information der Deutschen Nationalbibliothek:

Die Deutsche Bibliothek verzeichnet diese Publikation in der Deutschen Nationalbibliografie; detaillierte bibliografische Daten sind im Internet über http://dnb.d-nb.de/ abrufbar.

Dieses Werk sowie alle darin enthaltenen einzelnen Beiträge und Abbildungen sind urheberrechtlich geschützt. Jede Verwertung, die nicht ausdrücklich vom Urheberrechtsschutz zugelassen ist, bedarf der vorherigen Zustimmung des Verlages. Das gilt insbesondere für Vervielfältigungen, Bearbeitungen, Übersetzungen, Mikroverfilmungen, Auswertungen durch Datenbanken und für die Einspeicherung und Verarbeitung in elektronische Systeme. Alle Rechte, auch die des auszugsweisen Nachdrucks, der fotomechanischen Wiedergabe (einschließlich Mikrokopie) sowie der Auswertung durch Datenbanken oder ähnliche Einrichtungen, vorbehalten.

Imprint:

Copyright © 2012 GRIN Verlag GmbH
Druck und Bindung: Books on Demand GmbH, Norderstedt Germany
ISBN: 978-3-656-34647-0

This book at GRIN:

http://www.grin.com/en/e-book/203837/effect-of-carbohydrates-and-amino-acids-on-fermentative-production-of-alpha

GRIN - Your knowledge has value

Der GRIN Verlag publiziert seit 1998 wissenschaftliche Arbeiten von Studenten, Hochschullehrern und anderen Akademikern als eBook und gedrucktes Buch. Die Verlagswebsite www.grin.com ist die ideale Plattform zur Veröffentlichung von Hausarbeiten, Abschlussarbeiten, wissenschaftlichen Aufsätzen, Dissertationen und Fachbüchern.

Visit us on the internet:

http://www.grin.com/

http://www.facebook.com/grincom

http://www.twitter.com/grin_com

Effect of various carbohydrates and amino acids on production of fungal alpha amylase by solid state fermentation utilizing agricultural wastes

Debajyoti Bose[1] and Hiranmay Gangopadhyay[2]

Amity Institute of Biotechnology, Amity University Rajasthan, Jaipur-303002, India[1]

Department of Food Technology & Biochemical Engineering, Jadavpur University, Kolkata-700032, India[2]

Abstract

Extracellular alpha amylase was produced from *Aspergillus oryzae* under solid state fermentation. House hold agro-wastes were used as medium which were considered as one of the major pollutants due to unfavorable gas production via natural fermentation beside creating disposal problem. Investigations were carried out to evaluate the effect of various carbohydrate sources e.g. glucose, maltose, lactose, sucrose and soluble starch with different concentrations on the production of fungal alpha amylase utilizing agricultural wastes as the fermentation medium. Studies were also carried out to evaluate the effect of various heat stable amino acids e.g. glycine, histidine, proline, leucine and isoleucine with different concentrations on the production of fungal alpha amylase utilizing agricultural wastes as the fermentation medium. The results indicated that maximum activity of alpha amylase (6639.85 U/gds) was obtained at 0.5% concentration of sucrose solution with a r^2 of 0.9692 when compared to control (3319.95 U/gds) and other carbohydrates used. Maximum activity of alpha amylase (10407.80 U/gds) was also obtained at 0.75% concentration of proline solution with an r^2 of 0.6816 when compared to control (3035.60 U/gds) and other amino acids used in our present study.

Key words: alpha amylase, solid state fermentation, *Aspergillus oryzae*, agricultural wastes, carbohydrates, amino acids.

Introduction

Alpha Amylases (E.C. 3.2.1.1.) are starch-degrading enzymes that catalyze the hydrolysis of internal α-1,4-O-glycosidic bonds in polysaccharides with the retention of α-anomeric configuration in the products. Most of the α-amylases are metalloenzymes, which require calcium ions (Ca^{2+}) for their activity, structural integrity and stability. They belong to family 13 (GH-13) of the glycoside hydrolase group of enzymes[1]. Amylases are one of the most important industrial enzymes that have a wide variety of applications ranging from conversion of starch to sugar syrups, to the production of cyclodextrins for the pharmaceutical industry. These enzymes account for about 30 % of the world's enzyme production[2]. The α-amylase family can roughly be divided into two groups: the starch hydrolyzing enzymes and the starch modifying, or transglycosylating enzymes. The enzymatic hydrolysis is preferred to acid hydrolysis in starch processing industry due to a number of advantages such as specificity of the reaction, stability of the generated products, lower energy requirements and elimination of neutralization steps[3]. Alpha amylase is produced by a variety of plants, animals, and microorganisms[4]. Some α-amylase producing fungi are from the genera *Aspergillus, Penicillium, Cephalosporium, Mucor, Candida, Neurospora,* and *Rhizopus*[5].

Agricultural waste is composed of organic wastes (animal excreta in the form of slurries and farmyard manures, spent mushroom compost, soiled water and silage effluent) and waste such as plastic, scrap machinery, fencing, pesticides, waste oils and veterinary medicines. There are a number of methods used to treat agricultural waste. These include spreading the waste on land under strict conditions, anaerobic digestion and composting. Residues from fruit and vegetable processing units are also considered as agricultural wastes. These wastes are one of the major causes for environmental pollution. In general most of this "wastes" may be used as cattle feed or converted to biogas or compost. But, greater environmental and economic benefits could result from the conversion of these by-products of higher value. This can be achieved either by using such materials as multifunctional food ingredient or in order to other processes within the concept of low-residue food production. Thus bio-conversion of these wastes not only reduces disposal problem but also environmental pollution along with production of value added products[6]. Fourteen different agro-residues were screened for alpha amylase production using *Bacillus amyloliquefaciens* ATCC 23842, where wheat bran (WB) and groundnut oil cake (GOC) in mass ratio of 1:1 was proved as the best substrate source[7].

Solid-state fermentation (SSF) is an alternative to the submerged fermentation (Smf). A closer evaluation of these two processes in recent years in several research centers throughout the world has revealed the enormous economical and practical advantages of SSF over SmF. These include non-aseptic conditions, use of raw materials as substrates, use of a wide variety of matrices, low capital cost, low energy expenditure, less expensive downstream processing, less water usage, lower wastewater output, potential higher volumetric productivity, higher concentration of the products, high reproducibility, lesser fermentation space, easier control of contamination, and generally simpler fermentation media[8,9,10,11]. This, however, does not mean that SSF can be taken as a foolproof technology to replace SmF. SSF too has some disadvantages, such as: difficulty in agitation of the substrate bed, resulting in heterogeneously distributed physiological, physical and chemical environment in the substrate bed; difficulties in fermentation control, mainly in heat buildup; control of moisture level of the substrate and control of aeration; difficulty in rapid determination of microbial growth and other fermentative parameters; limited types of microorganisms that can grow at low moisture levels[8,9,12]. Wheat bran (WB) showed the highest enzyme production under optimum conditions of solid-state fermentation (SSF) carried out using corncob leaf (CL), rye straw (RS),wheat straw (WS) and wheat bran (WB) as substrates by a fungal culture of *Penicillium chrysogenum*[13].

Carbohydrates such as maltose, starch, cellobiose, lactose, glucose, fructose and galactose favours induction of amyloglucosidase(glucoamylase)[14]. Experimental studies had revealed that, different carbon sources such as glucose, sucrose, maltose, lactose and starch at 1% (w/w) concentration when mixed with substrate, shows an increased production of the alpha amylase enzyme, especially in cases with lactose and maltose, even with different strains of *Aspergillus oryzae*[15]. Starch and fructose may also serve as potential activators in case of amylase synthesis[16]. Amylase production was found to be stimulated due to fructose, lactose and maltose in all five species of *Fusarium*(*F. dimerum, F. moniliforme, F. oxysporum, F. roseum* and *F.semitectum*) while, Sucrose inhibited *F. oxysporum*[17]. Alpha amylase was also produced under solid-state fermentation by *Bacillus cereus* MTCC 1305 using wheat bran and rice flake manufacturing waste as substrates where glucose showed enhanced enzyme production, whereas supplementation of different nitrogen sources showed decline in enzyme production[18].

Amino acids such alanine, arginine, glycine, leucine, phenylalanine, proline, and cystine can act as stimulator for the production of alpha amylase from microorganisms[19] . Whereas, glycine, lysine, isoleucine and histidine, had proved to be vital for glucoamylase synthesis from *Aspergillus sp*. using rice bran as fermentation medium[14].

Considering the above factors, studies were carried out to evaluate the effect of various carbohydrates e.g. glucose, maltose, lactose, sucrose and soluble starch and heat sterilizable amino acids e.g. glycine, histidine, proline, leucine and isoleucine with different concentrations on the production of fungal alpha amylase utilizing agricultural wastes as the fermentation medium.

Materials & Methods

Microorganism

Aspergillus oryzae (NCIM No. 645) collected from National Collection of Industrial Microorganisms (NCIM), National Chemical Laboratory, Pune(India) was maintained on Czapek Dox agar medium and stored under refrigerated condition at 4°C. A suspension of the mold i.e. one loopfull in 5ml of sterile water blank was used as the inoculum for each Roux bottle in our present study. A constant ratio of 4:1 (w/v) of waste to inoculum was maintained through out the study.

Utilization of agricultural wastes for SSF

House hold agro-wastes i.e. waste part of vegetables (potato, peas and spinach) in Kolkata Municipal area were used as substrate in this study. These agro-wastes were sun dried for about 2 days and then dried at 60°C in Tray drier for 4 hr. and made to powder in a Mixer grinder. These agro-waste powder was used as medium for SSF through out the study. The mixed wastes in our present study showed initial moisture of 10%.

Preparation of Carbohydrate Solutions

Production of enzyme alpha-amylase by *Aspergillus species* varies with the addition of different carbohydrates acting as carbon sources with different concentrations. Carbohydrate solutions were prepared in distilled water with concentration of. 0.1%, 0.2%, 0.25%, 0.5% & 1%.The carbohydrates used in our present study as carbon sources are glucose, maltose, lactose, sucrose and soluble starch.

Preparation of amino acid Solutions

Production of enzyme alpha-amylase by *Aspergillus species* varies with the addition of different amino acids acting as nitrogen sources with different concentrations. Amino acid solutions were prepared in double distilled water with concentration of. 0.25%, 0.5%, 0.75% & 1%.The amino acids used as nitrtogen sources are glycine, histidine, proline, leucine and isoleucine.

Production of alpha amylase by SSF

Production of alpha amylase by *Aspergillus oryzae* was carried out in two sets of standard size Roux bottle(1L capacity). In one set 20g of agro-waste material (particle size of 0.48 mm) & 20ml of carbohydrate solutions 1:1 (w/v) with desired concentration was taken. In another set 20g of agro-waste material with same particle size & 20ml of amino acid solutions 1:1 (w/v) with desired concentration was taken. Finally, after aseptic inoculation with *Aspergillus oryzae*, both the sets were plugged with cotton wool aseptically. The fermentation was carried out under stationary condition at 30±2°C. Alpha amylase secreted into the spent medium was monitored at regular interval of time. After 5th day of incubation, both sets of Roux bottles were removed. The enzyme was extracted with distilled water and double distilled water by shaking for 4 hrs. at 30±2°C for the medium treated with carbohydrate and amino acid respectively. The ratio of waste to water was 1:2.5 w/v. Solid were removed by filtration followed by centrifugation at 10,000 rpm (C-24, REMI, India) for 20 minutes. Clear supernatants were used for measurement of alpha amylase activity.

Required environmental conditions for fermentation were optimized previously for the alpha-amylase produced[20]. A control was maintained for each set by adding only distilled water and double distilled water respectively instead of above mentioned carbohydrate and amino acid solutions.

Enzyme assay

Activity of enzyme produced was measured in Units (U) for alpha amylase. One unit of alpha amylase activity is defined as the amount of enzyme that releases 1μmole (micromole) of reducing sugar per minute from soluble starch at pH 7.0 and 30±2°C.Alpha amylase activity was determined at 30±2°C by mixing enzyme solution with 1.1% soluble starch(Merck Ltd.) dissolved in 0.05M Imidazole-HCl buffer, pH 7.0[21,22].

The enzyme activity for above mentioned enzyme was expressed in U/gds (i.e. gram dry solid) by taking the dry weight of the fermented solid agro-waste after drying it in 100°C-105°C for 12 hours (overnight) in a tray dryer[21].

Results & Discussion

Each set of experiment was performed in triplicates and the average values obtained for the enzyme activities were considered in our present study.

From the result it is clear that, alpha amylase showed a maximum activity of 5643.85 U/gds at 1% concentration of glucose (Fig. 1) which also showed an exponential increase in enzymatic activity, 5809.90 U/gds at 0.25% concentration of maltose (Fig. 2), 1778.53 U/gds at 0.1% concentration of lactose (Fig. 3) along with an exponential decrease in enzymatic activity, 6639.85 U/gds at 0.5% concentration of sucrose (Fig. 4), 4387.06 U/gds at 0.1% concentration of soluble starch (Fig. 5) with a control (without carbohydrate solution) showing an activity of 3319.95 U/gds. Increase yield after incorporation of additional carbon source during SSF was reported by M.E.Upton and W.M.Fogarty (1977)[23]; B.K.Gogoi et al., (1987)[24]; H.Melasneimi (1987)[25]; S.N.Freer (1993)[26]; L.L.Lin et al., (1998)[27] and G.Mamo and A.Gessesse (1999)[28]. The results obtained in our experiment were also supported by the data obtained from an experiment to observe the effect of 1%(w/v) glucose and maltose solutions on amylase production, conducted by P.V.D.Aiyer (2004)[16]. Alpha amylase is an inducible enzyme and is generally induced in the presence of starch or its hydrolytic product, maltose as repoted by K.Tonomura et al.(1961)[29]; M.Yabuki et al.(1977)[30]; A.Lachmund et al. (1993)[31]; R.Morkeberg et al.(1995)[32].Most reports available on the induction of alpha amylase in different strains of *Aspergillus oryzae* suggest that the general inducer molecule is maltose. Apart from maltose, in some strains, other carbon sources such as lactose, trehalose, alpha-methyl-D-glucosidase also served as inducers of alpha amylase as reported by M.Yabuki et al. (1977)[30].

Alpha amylase production is also subjected to catabolite repression by glucose and other sugars, like most other inducible enzymes as per the report of R.S.Bhella and I.Altosaar (1985)[33]; R.Morkeberg et al. (1995)[32].However, the role of glucose in the production of alpha amylase in certain case is controversial. Alpha amylase production by *Aspergillus oryzae* DSM 63303 was not repressed by glucose rather, a minimum level of enzyme was induced in its presence as per the report of A.Lachmund et al., (1993)[31], which intern also supports the results obtained using glucose in our present study. S.N.Freer (1993)[26] reported that in *Streptococcus bovis* JB1, starch and maltose induced 12 fold amylase activities than glucose as sole carbon source. A possible reason for the result obtained in our present study may be due to the necessity of some carbon sources in specific concentrations for the metabolism of the microorganisms, which intern enhances the production and activity of the enzyme, where as other carbon sources may suppress the activity of the above said enzyme. An experiment performed by S.Ramachandran et al. (2004)[21] showed an increased enzyme activity with the increase in starch concentration at 1%, after which the activity remained stable. Whereas, an enzyme activating effect at lower concentration and inhibitory effect at higher concentration of soluble starch (Fig.5) were observed in our present study, which may open some new ideas and need further investigations at molecular level to demonstrate a proper reason.

From the result it is clear that, alpha amylase showed a maximum activity of 4336.50 U/gds at 1% concentration of glycine (Fig. 6), 9540.40 U/gds at 0.5% concentration of histidine (Fig. 7), 10407.80 U/gds at 0.75% concentration of proline (Fig. 8), 4553.30 U/gds at 0.25% concentration of leucine (Fig. 9) and 4444.90 U/gds at 0.5% concentration of isoleucine (Fig. 10) with a control (without amino acid solution) showing an activity of 3035.60 U/gds. The role of amino acids was considered nither as nitrogen or as a carbon source, but as stimulators of amylase synthesis and excretion by Y.Ikura and K.Horikoshi (1987)[34]. Q.Zhang et al. (1983)[35] reported that alpha amylase production by *Bacillus amyloliquefaciens* ATCC 23350 was increased by a factor of 300 in presence of glycine. Glycine not only acted as a source of nitrogen but rather it affected alpha amylase production by controlling pH thereby increasing amylase production. The results obtained in our experiment especially with glycine, leucine and proline were also supported by the data obtained from the experiment performed by S.Aguloglu et al (2000)[19]. In another experiment conducted by S.Ali et al (1989)[14] on glucoamylase synthesis, data obtained with isolucine and histidine supported the results of our present study. A possible reason for this type of result may be due to the necessity of some nitrogen sources in specific concentrations, also acting as the growth factor for the microorganisms, which intern enhances the production and activity of the enzyme during metabolism. When a concentration of 0.25% of glycine was used in our present study (Fig.1), an enzyme inhibitory effect were observed which may open some new ideas of low concentration inhibition capacity of amino acids and need more study to detect a proper reason. Fig. 1 also showed an exponential pattern for increase in enzyme activity possible due to direct proportionality between concentration of glycine and enzyme activity.

All the above results were properly analyzed by appropriate statistical tools [36].

Conclusion

Except lactose all other carbohydrate sources used in our experiment gave an increased activity at specific concentrations for the enzyme alpha amylase compared to the activity shown by the control, whereas all the amino acids used in our experiment gave an increased activity for the enzyme alpha amylase compared to the activity shown by the control at a specific concentration. The increase in enzyme activities was directly proportional with concentration of amino acids in some cases. But, they were maximum with specific concentrations. This also signifies the novelty of the research work where the agricultural wastes which are considered as unwanted and discarded materials of our society, were utilized to produce higher amount of commercially valuable product (enzymes) by making changes in the nutritional requirement, especially some carbohydrates and amino acid sources with specific concentration for the mold, exploited in an eco-friendly fermentation process.

Acknowledgement

The authors are highly grateful to University Grants Commission (UGC), New Delhi for providing the financial assistance.

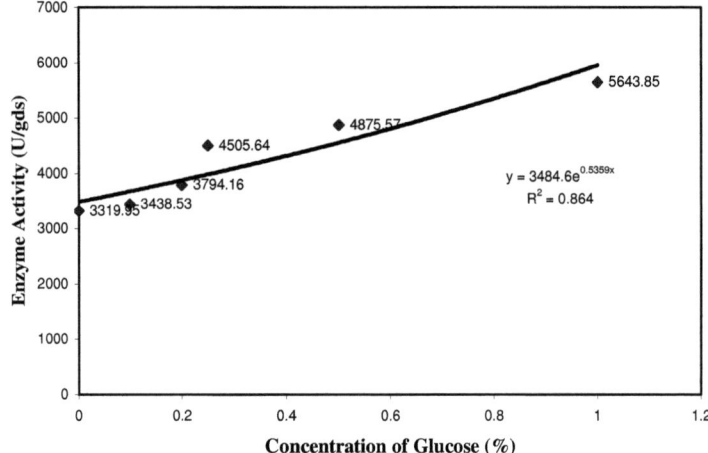

Fig. 1

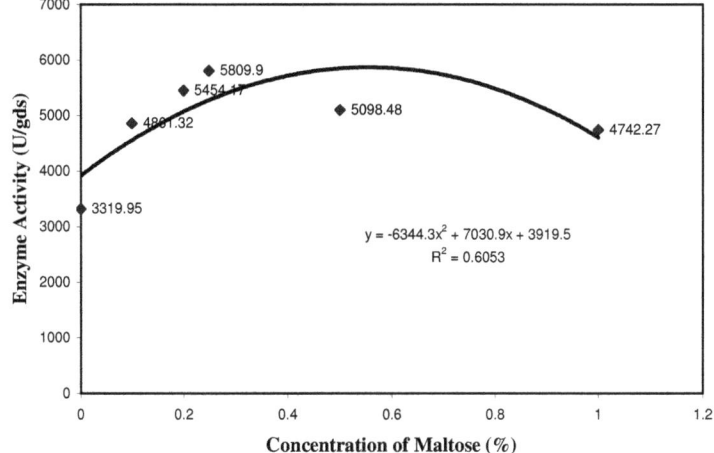

Fig. 2

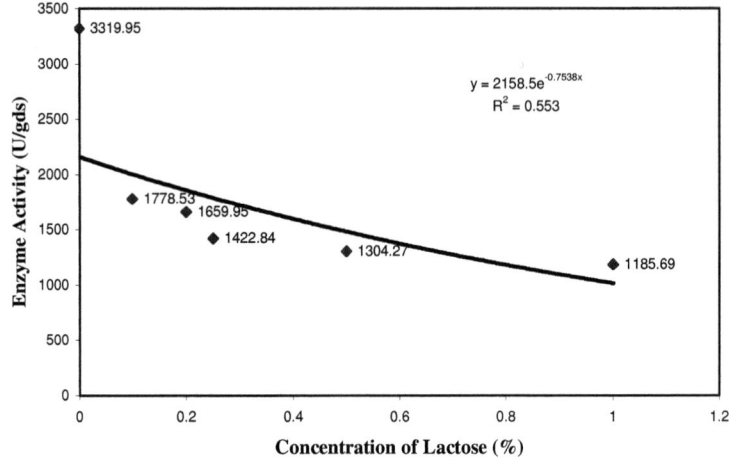

Fig. 3

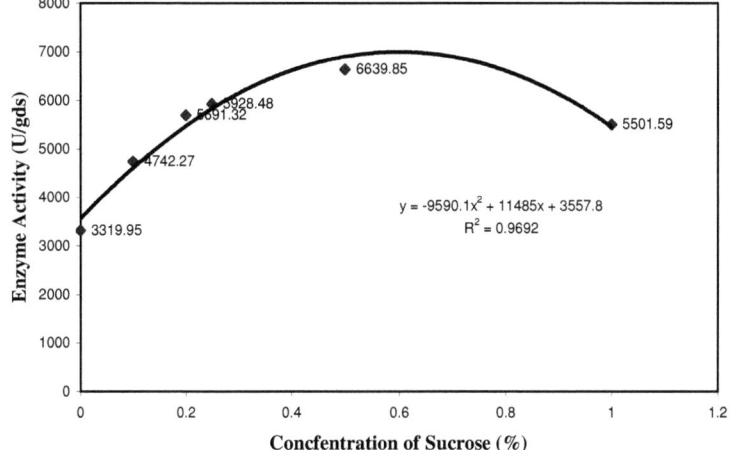

Fig. 4

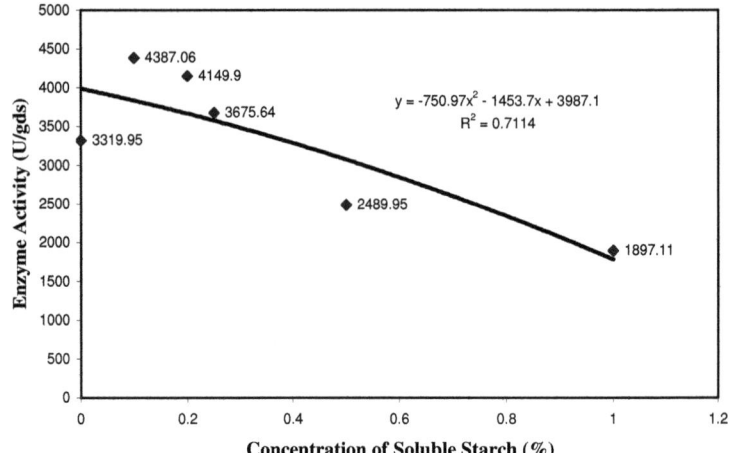

Fig. 5

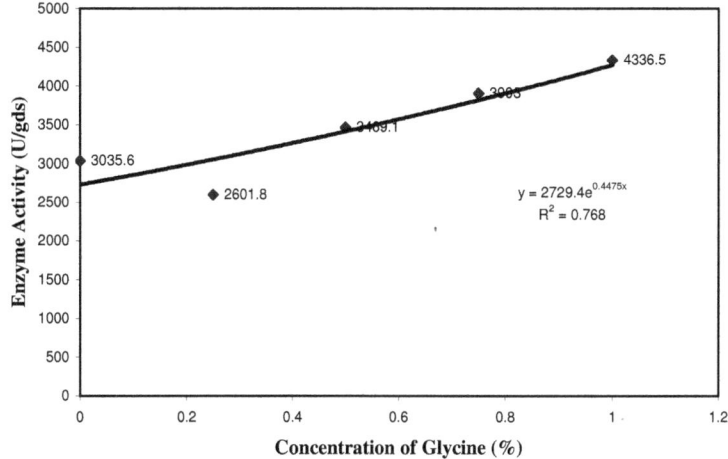

Fig. 6

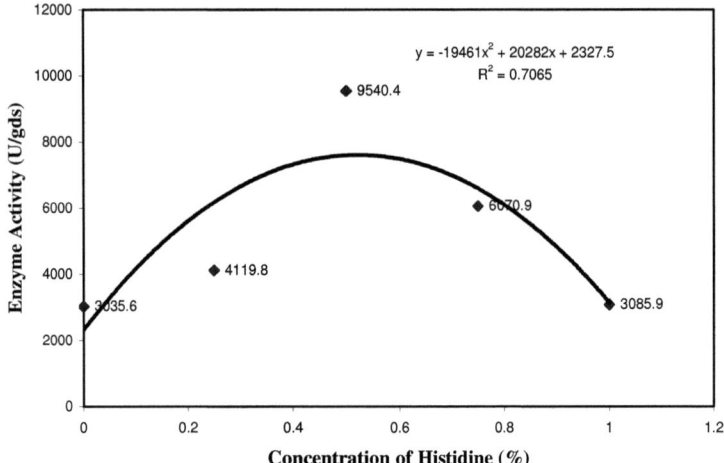

Fig. 7

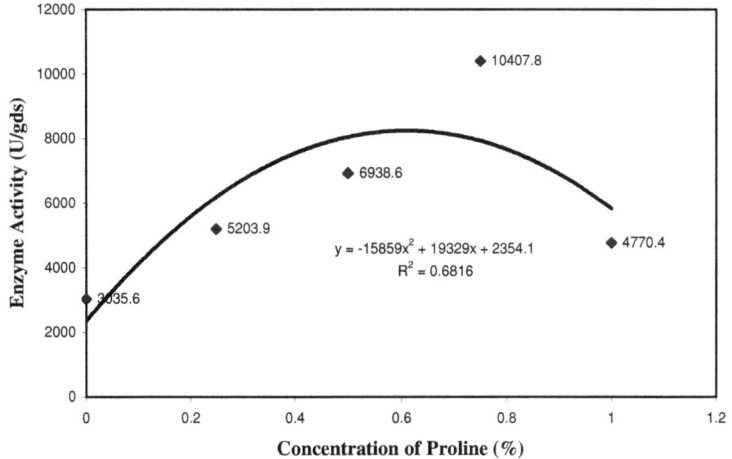

Fig. 8

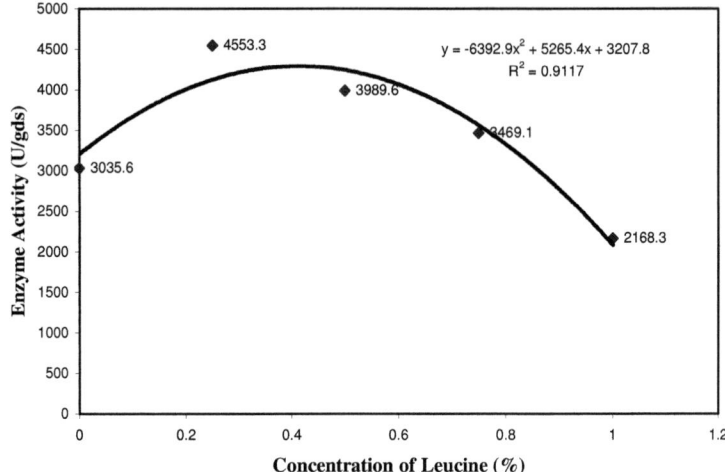

Fig. 9

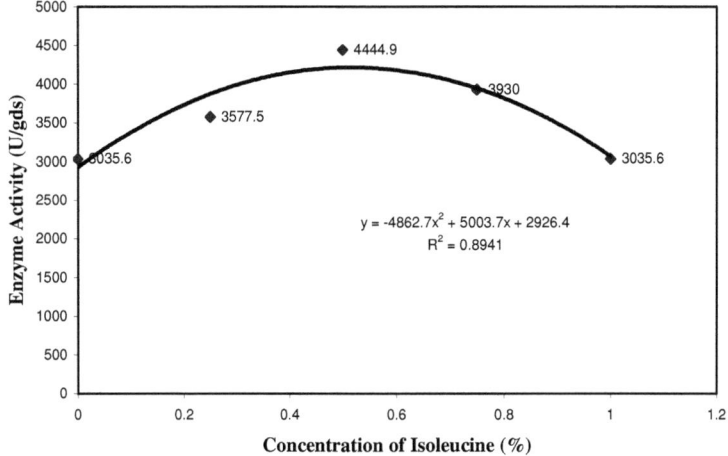

Fig. 10

References

[1] A.K. Bordbar, K. Omidiyan and R. Hosseinzadeh (2005) Study on interaction of a-amylase from *Bacillus subtilis* with cetyl trimethylammonium bromide. *Colloids Surf. B: Biointerfaces.* 40:67–71.

[2] M.J.E.C. van der Maarel, B. van der Veen, J.C.M. Uitdehaag, H. Leemhuis and L. Dijkhuizen (2002) Properties and applications of starch-converting enzymes of the α-amylase family. *J. Biotechnol.* 94:137–155.

[3] T. Satyanarayana, J.L.U.M. Rao and M. Ezhilvannan (2005) α-Amylases. pp. 189–220. In: A. Pandey, C. Webb, C.R. Soccol and C. Larroche (eds.). *Enzyme Technology.* Asiatech Publishers Inc., New Delhi, India.

[4] A.Pandey, C.R.Soccol, J.A.Rodriguez-Leon and P.Nigam (2001) *Solid-State Fermentation in Biotechnology—Fundamentals and Applications.* pp. 100-221. Asiatech Publishers Inc, New Delhi, India.

[5] F.N.Yasouri and R.Sariri (2003) Extraction and Purification of a Thermostable α-Amylase from *Bacillus Coagulans* found in Iranian Soil Samples. *Int. J.Chem. Sc.*1: 332-340.

[6] D.Bose D, U.Ghosh and H.Gangopadhyay (2004) Production of amyloglucosidase enzyme from agricultural wastes by solid state fermentation using *Aspergillus oryzae*. *Biohorizon2004, 6th National Symposium on Biochemical Engineering and Biotechnology. Department of Biochemical Engineering & Biotechnology.* IIT Delhi, India.

[7] D.Gnadharan, S.Sivaramakrishnan, K.M.Nampoothiri and A.Pandey (2006) α-Amylase Production by *B. amyloliquefaciens. Food Technol. Biotechnol.* 44:269–274.

[8] M.K.Gowthaman, C.Krishna and M.Moo-Young (2001) Fungal solid state fermentation—An overview in Applied Mycology and Biotechnology. pp.305-352. In: G.G.Khachatourians and D.K.Arora (eds.). *Agriculture and Food Productions.* Elsevier Science, The Netherlands.

[9] K.Sato and S.Sudo (1999) Small scale solid state fermentations.pp.61-79. In: A.L.Demain and J.E.Davies (eds.). *Manual of Industrial Microbiology and Biotechnology*. ASM Press, Washington DC.

[10] A.Durand and D.A.Chereau (1988) A new pilot reactor for solid state fermentation: Application to the protein enrichment of sugar beet pulp. *Biotechnol. Bioeng.* 31: 486– 476.

[11] A.Durand (2003) Bioreactor designs for solid state fermentation. *Biochem.Eng. J.*13:113– 125.

[12] U.Holker, M.Hofer and J.Lenz (2004) Biotechnological advantages of laboratory-scale solid-state fermentation with fungi. *Appl. Microbiol. Biotech.* 64:175–186.

[13] B.Balkan and F.Ertan (2007) Production of α-Amylase from *P. chrysogenum*. *Food Technol. Biotechnol.* 45:439–442.

[14] S.Ali, S.Mahmood, R.Alam and Z.Hossain (1989) Culture condition for production of glucoamylase from rice bran by *Aspergillus terreus*. *World. J. Microbio. Biotech.* 5:525-532.

[15] S.Ramachandran, A.K.Patel, K.M.Nampoothiri, S.Chandran and G.Szakacs (2004) Alpha Amylase from a Fungal Culture Grown on Oil Cakes and Its Properties. *Brazilian. Arc. Bio. Tech.* 47:309-317.

[16] P.V.D.Aiyer. Effect of C:N ratio on alpha amylase production by *Bacillus licheniformis* SPT 27. *African. J. Biotech.* 3:519-522.

[17] B.M.Waghmare, S.R.Shinde, G.T.Sumanth and V.T.Gorgile (2010) Effect of Carbohydrates on production of hydrolytic enzymes in different species of *Fusarium*. *International Journal of Plant Sciences*.5 : 577-578.

[18] H.Anto, H.Trivedi and K.Patel (2006) α-Amylase Production by *B. cereus*. *Food Technol. Biotechnol.* 44:241–245.

[19] S.Aguloglu, N.Y.Ensari, F.Uyar and B.Otludil (2000) The Effects of Amino Acids on Production and Transport of α–Amylase through Bacterial Membranes. *Starch.Strake*.52:290-295.

[20] D.Bose, U.Ghosh and H.Gangopadhyay (2009) Production of extracellular alpha amylase from *Aspergillus oryzae* by solid state fermentation utilizing agricultural wastes. *J. Micopathol. Res*.47:153-159.

[21] J.F.Shaw, F.P.Lin, S.C.Chen and H.C.Chen (1995) Purification and properties of an extracellular α-amylase from *Thermus sp. Bot. Bull. Acad. Sin*. 36:195-200.

[22] G.L.Miller (1959) Use of Dinitrosalicylic Acid Reagent for Determination of Reducing Sugar. *Anal. Chem*.31:426-428.

[23] M.E.Upton and W.M.Fogarty (1977) Production and Purification of Thermostable Amylase and Protease of *Thermomonospora viridis*. *Appl. Environ. Microbiol*. 33:59-64.

[24] B.K.Gogoi, R.L.Bezbaruah, K.R.Pillai and J.N.Baruah (1987) Production, purification and characterization of an α-amylase produced by *Saccharomycopsis fibuligera*. *J.Appl. Microbiol*. 63:373-379.

[25] H.Melasneimi (1987) Effect of Carbon Source on Production of Thermostable α-Amylase, Pullulanase and α-Glucosidase by *Clostridium Thermohydrosulfuricum*. *J. Gen. Microbiol*.133:883-890.

[26]S.N. Freer (1993) Purification and Characterization of the Extracellular α -Amylase from *Streptococcus bovis* JB1. *Appl. Environ. Microbiol*. 59:1398-1402.

[27] L.L.Lin, C.C.Chyau and W.H.Hsu (1998) Production and properties of a raw-starch degrading amylase from the thermophilic and alkaliphilic *Bacillus* sp. TS-23. *Biotechnol. Appl. Biochem*. 28:61-68.

[28] G.Mamo and A.Gessesse (1999) Effect of cultivation conditions on growth and α-amylase production by a thermophilic *Bacillus* sp. *Letters. Appl. Microbiol*. 21:61-65.

[29] K.Tonomura, H.Suzuki, N.Nakamura, K.Kuraya and O.Tanabe (1961) On the inducers of α-amylase formation in *Aspergillus oryzae*. *Agri Biol. Chem.* 25:1-6.

[30] M.Yabuki, N.Ono, K.Hoshino and S.Fukui (1977) Rapid Induction of α-Amylase by Nongrowing Mycelia of *Aspergillus oryzae*. *Appl. Environ. Microbiol.* 34:1-6.

[31] A.Lachmund, U.Urmann, K.Minol, S.Wirsel and E.Ruttkowski (1993) Regulation of α-amylase formation in *Aspergillus oryzae* and *Aspergillus nidulans* transformants. *Curr. Microbiol.* 26:47-51.

[32] R.Morkeberg, M.Carlsen and J. Nielsen (1995) Induction and repression of α-amylase production in batch and continuous cultures of *Aspergillus oryzae*. *Microbio.* 141:2449-2454.

[33] R.S.Bhella and I.Altosaar (1985) Purification and some properties of the extracellular α-amylase from *Aspergillus awamori*. *Can. J. Microbiol.* 31:149-153.

[34] Y.Ikura and K.Horikoshi (1987) Effect of amino compounds on alkaline amylase production by alkalophilic *Bacillus* sp. *J. Ferment. Technol.* 65:707-709.

[35] Q.Zhang, N.Tsukagoshi, S.Miyashiro and S.Udaka. Increased production of a-amylase by *Bacillus amyloliquefaciens* in the presence of glycine. *Appl. Environ. Microbiol.* 46:293-295.

[36] R.R.Sokal and F.J.Rohlf (1981) Biometry-The Priciple and Practice of Statistics in Biological Research. 2nd ed., pp.496. Freeman & Co., N Y, USA.

Figure Captions:

1. Fig.1 : Effect of glucose on production of alpha amylase from *Aspergillus oryzae* utilizing agricultural wastes.

2. Fig.2 : Effect of maltose on production of alpha amylase from *Aspergillus oryzae* utilizing agricultural wastes.

3. Fig.3 : Effect of lactose on production of alpha amylase from *Aspergillus oryzae* utilizing agricultural wastes.

4. Fig.4 : Effect of sucrose on production of alpha amylase from *Aspergillus oryzae* utilizing agricultural wastes.

5. Fig.5 : Effect of soluble starch on production of alpha amylase from *Aspergillus oryzae* utilizing agricultural wastes.

6. Fig.6 : Effect of glycine on production of alpha amylase from *Aspergillus oryzae* utilizing agricultural wastes.

7. Fig.7 : Effect of histidine on production of alpha amylase from *Aspergillus oryzae* utilizing agricultural wastes.

8. Fig.8 : Effect of proline on production of alpha amylase from *Aspergillus oryzae* utilizing agricultural wastes.

9. Fig.9 : Effect of leucine on production of alpha amylase from *Aspergillus oryzae* utilizing agricultural wastes.

10. Fig.10: Effect of isoleucine on production of alpha amylase from *Aspergillus oryzae* utilizing agricultural wastes.